Smoky Mountain Mystery

by Sharon Kahkonen

Orlando Austin New York San Diego Toronto London

Visit *The Learning Site!*
www.harcourtschool.com

Camping in the Mountains

Last July 4, my family and I set out on a backpacking trip in Great Smoky Mountains National Park. Little did I know that on that fateful day, I would begin my career as Sandra Brown, Nature Detective. This is a true story about how I solved the Smoky Mountain Mystery.

First, let me tell you a little about the mountains where all of this took place. The Smokies are among the oldest mountains in the world, formed some 200 to 300 million years ago. The park is home to more than 10,000 identified species. These mountains are called the Smokies because of the mist that hangs in the air like smoke. The air is very humid, and it rains a great deal. It sounds like the perfect setting for a mystery, don't you think?

We started our hike from a campground called Low Gap. With full packs, we hiked up to the Appalachian Trail, called

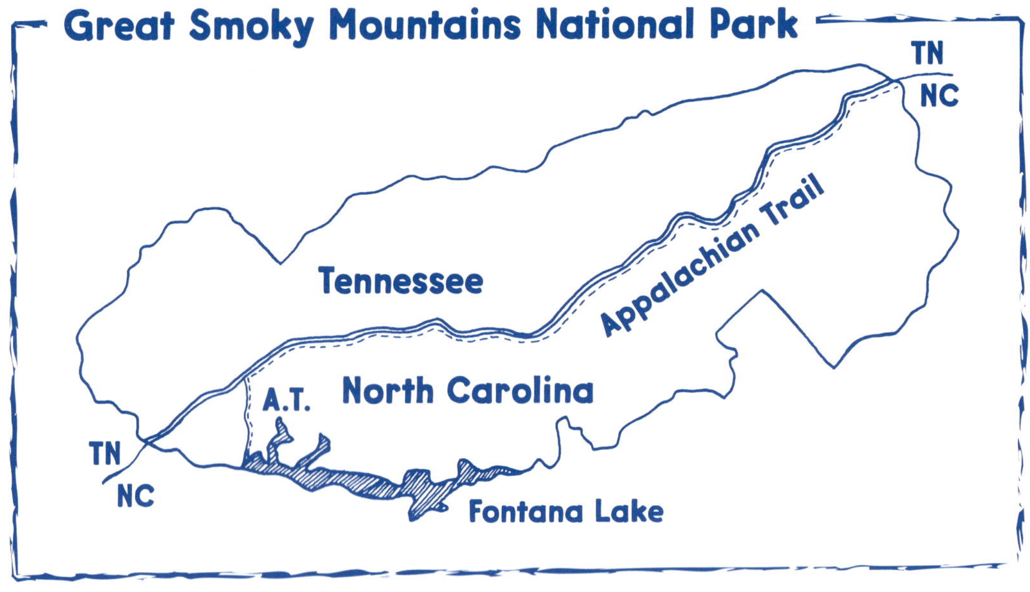

the AT for short. This trail runs from Georgia all the way to Maine. In the Smokies, the AT runs along the top ridge of the mountains. This area is called backcountry because it is far away from any roads or buildings. As we hiked, some white-tailed deer bounded across the trail in front of us. They were very well camouflaged. This means that their colors and markings made it hard for us to see them until they ran right in front of us.

Because it is so damp in the Smokies, the environment is perfect for salamanders. We found several different kinds just by digging through the leaf litter.

We had a long, hard trek to the top of the ridge, especially with all our gear on our backs, but the view from the top was spectacular! Our campsite was in a thick grove of trees next to a stream. We pitched our tents and cooked supper.

Nighttime Visitors

By the time we had finished eating, it was already getting dark. That's when we started to see small, dark shapes looping and fluttering above our heads—bats! Dad said the bats were our friends because they were snatching mosquitoes from the air. Still, they were a little spooky!

The bats were not our only visitors. About 9 meters (30 ft) away, we saw a skunk! We stayed at a safe distance, but we were able to get a closer look at it through our binoculars. It was tearing up a rotten log and eating the insects from under the bark. Skunks have a good sense of smell, but they can't see too well. They're not aggressive, so we knew that if we didn't bother this one, it wouldn't spray us.

Skunks are omnivores.

Skunks are omnivores. They eat everything—including the food in hikers' backpacks! Bears are omnivores, too. The park ranger had warned us to be sure to hang our backpacks from the trees, at least 3 meters (10 ft) from the ground. Otherwise, bears, skunks, or other animals might break into them and eat our food. We tied a rope from one tree to another about 5 meters (16 ft) away. Then we hung our backpacks from the middle of the rope.

After we had finished all of our preparations for the night, we sat around the campfire, and Dad told us a story. The last coals of our campfire died, and it began to rain, so we all turned in for the night. I curled up inside my warm sleeping bag, cozy and dry, and listened to the peaceful sound of raindrops on the tent.

Later that night, I woke up with a start. The rain had stopped. Outside our tent, there was a rustling in the leaves. It got closer and closer.

"What's that?" Tim whispered.

"It's probably just a little animal, maybe a mouse, looking for some food," I said, as calmly as I could.

Then we heard someone or something calling in an eerie voice, "Whoooo! Whoooo! Whooooo!" There was a whoosh and a pitiful cry. Then it was dead quiet again.

I grabbed a flashlight and aimed the beam into the trees, where the eerie sounds seemed to have come from. We got quite a fright when we saw two huge, shining eyes staring back at us! Our screams woke Mom and Dad. I shone the flashlight toward the trees again to show them what we had seen, but nothing was there. Dad told us that we were letting our imaginations run away with us.

We had other night visitors. I heard twigs snapping and a high-pitched chirring sound. Then something started splashing in the stream. It was definitely a lot larger than a mouse! I shone my flashlight toward the stream and saw one large pair of eyes and three smaller pairs of eyes, shining in the dark. There was a sharp, high-pitched bark and a lot of chattering. Then the eyes disappeared, and everything was quiet again.

By that time I was really tired and thought I could sleep through almost anything. Wrong! I don't think anyone could have slept through what happened next. First, there was a lot of growling, hissing, and barking. Then there was a scream that lasted for at least 30 seconds. We were all scared out of our wits! I decided right then that I was going to get to the bottom of this. As soon as it was daylight, I was going to find out who or what the night prowler had been.

Solving the Mystery

We woke at dawn and ate a light breakfast. Then I began looking for clues. The rain had left the ground muddy. Any animals that were around during the night should have left tracks. Sure enough, I found lots of tracks by the stream. Some looked like the handprints of tiny humans. Some were large—about 5 to 7 centimeters (2 to 3 in.) long, and others were smaller—only about 2 centimeters (1 in.) long. There were also pieces of crayfish near the stream.

I found animal tracks.

I showed the tracks to Mom and Dad and asked, "Could there be elves or fairies in these woods?" They laughed at the idea, but they did not know what animal might have made the humanlike tracks. Mom gave me a field guide that she had brought with her and suggested that I might find an answer in there. I looked through the guide, which had pictures of the tracks that each animal makes. I found what I was looking for—the tracks were definitely made by raccoons!

I was able to fit all of the pieces of the puzzle together. A mother raccoon and her babies had been catching crayfish in the stream. When we shone the flashlight on them, their eyes shone because they have a lining inside their eyes that reflects light. The chirping and chattering we heard was the mother warning her babies to escape to the safety of the trees.

It seemed obvious, now that it was daylight, that the two shining eyes in the tree had probably belonged to an owl. From the information in the field guide, I figured out that it was a Great Horned Owl that we heard. They have a call that sounds like "whooo-whooo-whoooo."

The Great Horned Owl rests in tree hollows during the day. I looked around the area and found a dead tree. At the base of the tree I found owl pellets. Owls swallow their food whole, but they can't digest fur, feathers, and bones. These waste materials are pressed into neat bundles by the owl's strong stomach muscles. Later the owl spits them out. If you take apart owl pellets, you can find out what the owl has been eating. Inside this pellet were plenty of bones—ribs, legs, vertebrae, and some tiny skulls.

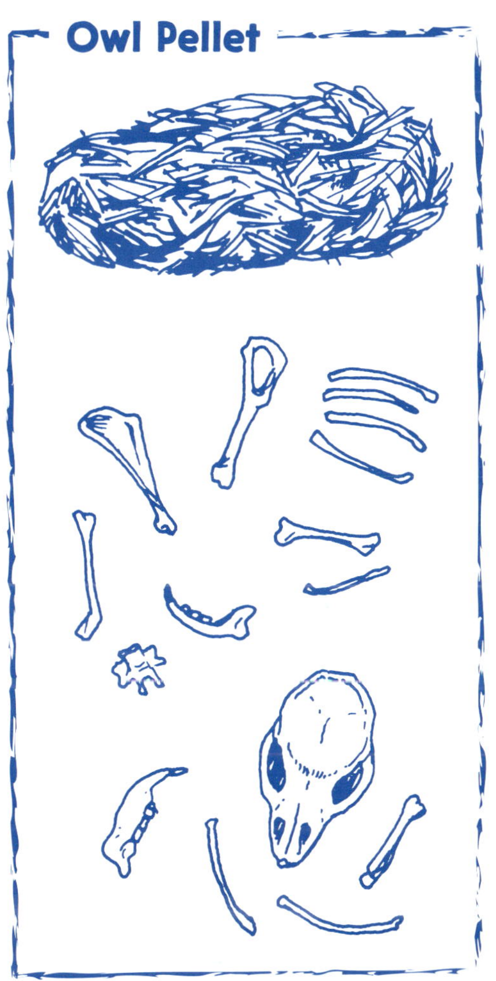

Who Screamed?

I felt satisfied with my solution to this part of the mystery. But what had made the terrible scream during the night? I looked around the area where I thought the scream had come from and found the tracks of two different animals.

These are fox tracks and opossum tracks.

The field guide showed that they were the tracks of an opossum and a fox. Fox tracks are dainty and make almost a straight line. Opossum tracks are star-shaped because their toes are spread far apart.

We had heard an encounter between a fox and an opossum. The fox had probably tried to attack the opossum. Foxes usually eat mice, and they are also fond of berries, fruits, and nuts. However, they sometimes attack larger animals, such as opossums. Usually an opossum will roll over and play dead when attacked, but sometimes it fights back. From the growling and hissing we heard, this opossum must have decided to fight. An opossum has very sharp teeth, and the fox must have gotten more than it bargained for and screamed. I knew that foxes barked, but I read that when a fox is hurt, it will also scream.

I was beginning to realize that these woods, which seemed to be so empty at first, were actually the habitat of many creatures. Some are not easily visible because they are so well camouflaged. Others sense us before we sense them and stay very still, or hide so that they cannot be seen. Nocturnal animals hunt for food after dark. That's when the woods around our campsite came alive with noises.

I began to think about how the living things in the Smokies depend on one another to survive. I made some sketches in my notebook. A good detective always carries a notebook. I wanted to record all of the interactions between animals that I had observed, so I decided to draw some food chains. I had learned in school that each food chain starts with a producer (a plant).

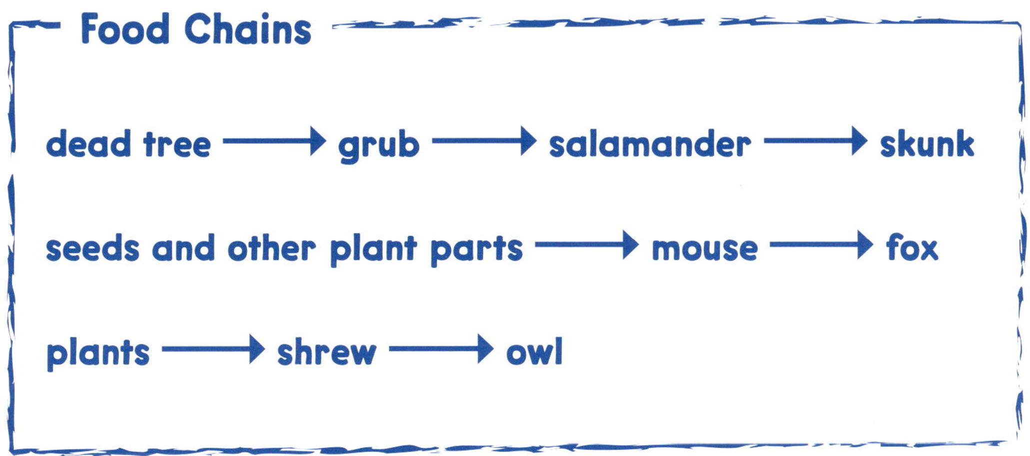

Of course, the real picture of how food energy moves through the Smoky Mountain community is more complicated than simple food chains. Most animals do not eat just one type of plant or one kind of animal. Food chains connect to form food webs. I drew a food web on the next page for the animals I observed in the Smokies.

My trip to the Smokies let me see for myself the community of animals and plants that live in there. My adventure was certainly much livelier than just reading about it would have been! It also gave me a start in my new career—Sandra Brown, Nature Detective. Look me up if you ever need help in solving a nature mystery!

Food web for animals in the Smokies

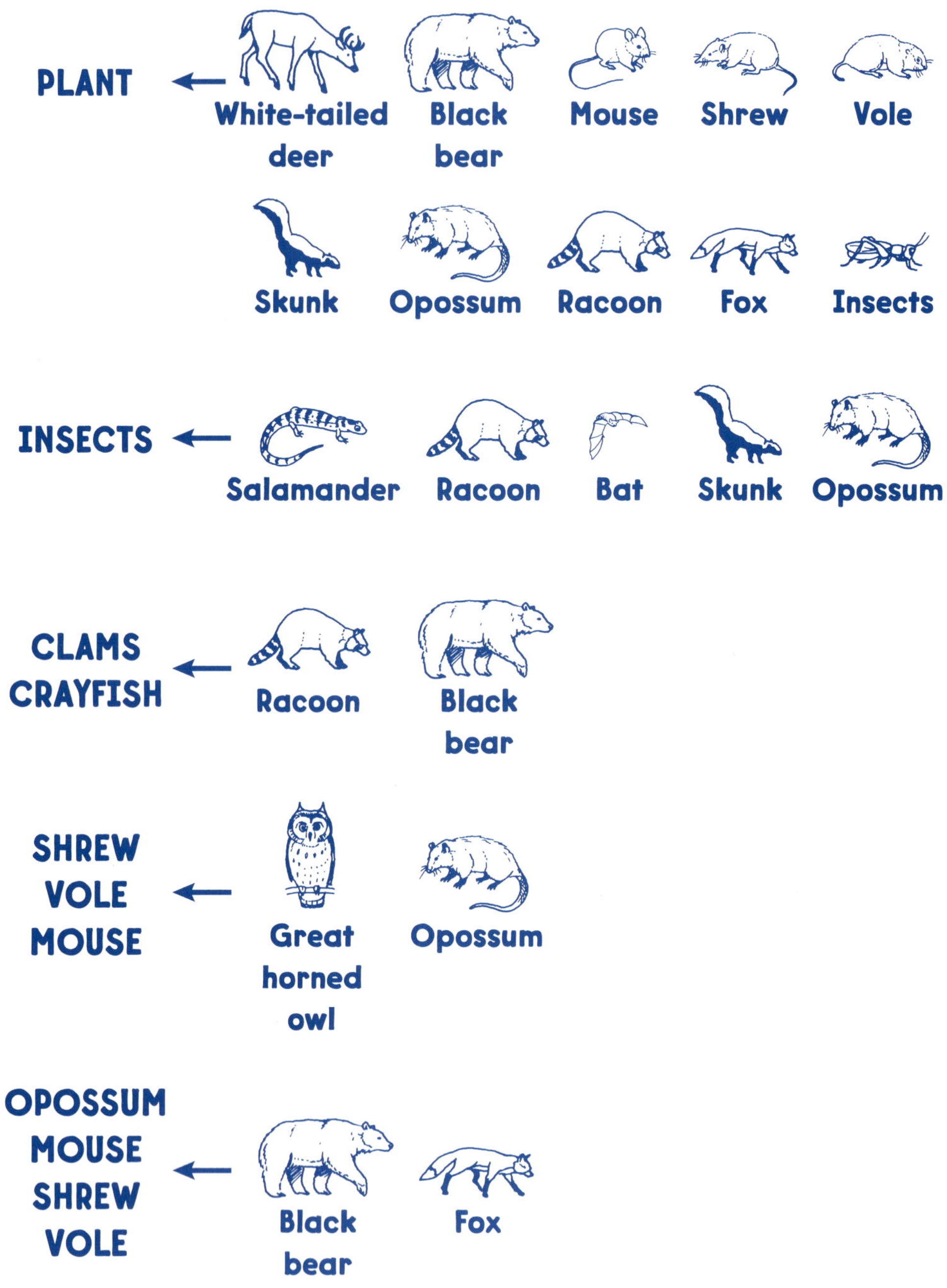

PLANT ←	White-tailed deer, Black bear, Mouse, Shrew, Vole, Skunk, Opossum, Racoon, Fox, Insects
INSECTS ←	Salamander, Racoon, Bat, Skunk, Opossum
CLAMS CRAYFISH ←	Racoon, Black bear
SHREW VOLE MOUSE ←	Great horned owl, Opossum
OPOSSUM MOUSE SHREW VOLE ←	Black bear, Fox